discovermore

A Look at Weather and Climate

Forecasting the Weather

Marie Harts

Britannica Educational Publishing

IN ASSOCIATION WITH

THE ROSEN PUBLISHING GROUP

Published in 2025 by Britannica Educational Publishing (a trademark of Encyclopædia Britannica, Inc.) in association with The Rosen Publishing Group, Inc.
2544 Clinton Street, Buffalo, NY 14224

Distributed exclusively by Rosen Publishing.
To see additional Britannica Educational Publishing titles, go to rosenpublishing.com.

Editor: Brianna Propis
Book Design: Michael Flynn

Photo Credits: Cover SpiffyJ/iStock; (series background) Dai Yim/Shutterstock.com; p. 5 (top) Gorodenkoff/Shutterstock.com; p. 5 (bottom) Image Point Fr/Shutterstock.com; p. 6 ZaZa Studio/Shutterstock.com; p. 7 Stock Lpa/Shutterstock.com; p. 9 Rafael Trafaniuc/Shutterstock.com; p. 10 Alexander_Evgenyevich/Shutterstock.com; p. 11 Taeya18/Shutterstock.com; p. 12 motive56/Shutterstock.com; p. 13 Broadcast Media/Shutterstock.com; p. 14 Dennis MacDonald/Shutterstock.com; p. 15 muratart/Shutterstock.com; p. 16 ocphoto/Shutterstock.com; p. 17 Karol Cygal/Shutterstock.com; p. 19 Gorodenkoff/Shutterstock.com; p. 20 Roman Mikhailiuk/Shutterstock.com; p. 21 Saigh Anees/Shutterstock.com; p. 22 Zenobillis/Shutterstock.com; p. 23 Nakantee/Shutterstock.com; p. 25 KingShopArt/Shutterstock.com; p. 26 J.J. Gouin/Shutterstock.com; p. 27 UweMosch/Shutterstock.com; p. 28 ND700/Shutterstock.com; p. 29 Tunatura/Shutterstock.com.

Cataloging-in-Publication Data

Names: Harts, Marie.
Title: Forecasting the weather / Marie Harts.
Description: New York : Britannica Educational Publishing, in Association with Rosen Educational Services, 2025. | Series: Discover more! a look at weather and climate | Includes glossary and index.
Identifiers: ISBN 9781641901321 (library bound) | ISBN 9781641901314 (pbk) | ISBN 9781641901338 (ebook)
Subjects: LCSH: Weather forecasting--Juvenile literature. | Meteorology--Juvenile literature.
Classification: LCC QC995.43 W347 2025 | DDC 551.5--dc23

Manufactured in the United States of America

Some of the images in this book illustrate individuals who are models. The depictions do not imply actual situations or events.

CPSIA Compliance Information: Batch #CSBRIT25. For further information contact Rosen Publishing at 1-800-237-9932.

Find us on

Contents

Shorts or Pants?

No matter where you are on Earth, the weather is always changing. Some patterns, like our seasons, are predictable. They come and go at the same times each year. Other patterns and events change often. Meteorology is a field of science that studies the changes in weather on a day-to-day basis in a specific place. Scientists who study meteorology are called meteorologists.

Using past history, patterns, and various measurement tools, meteorologists can tell whether it will rain or snow and whether it will be warm or cold. These scientists are able to tell people what the weather will be like anywhere in the world. This is called forecasting.

Climatology is the study of climate. Climate is the average weather conditions of a particular place or region over a period of years. Climatologists study various weather patterns over a long period of time. Their work is similar to meteorologists, but climatologists focus on a much longer amount of time.

Weather forecasts for the day and week are broadcast on television. Forecasters use maps to help people understand upcoming weather conditions.

compareandcontrast

How are meteorologists and climatologists different? How are they similar?

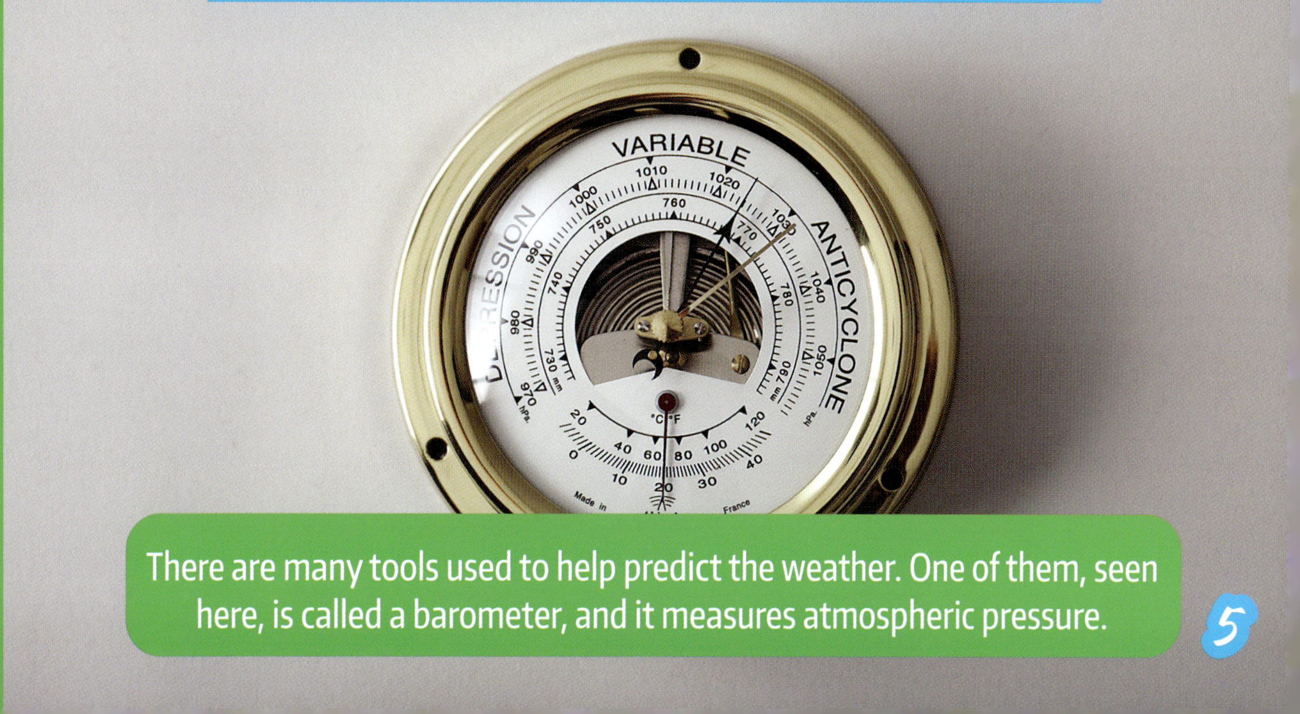

There are many tools used to help predict the weather. One of them, seen here, is called a barometer, and it measures atmospheric pressure.

Defining Weather

Weather is the state of the atmosphere in any given place at a specific time. The atmosphere is made up of the gases that surround Earth. Weather takes place in the lowest level of the atmosphere.

To most of us, the weather means things like it is hot and sunny outside or it is cold and rainy. To a meteorologist, the weather is a combination of certain features. A meteorologist can watch and measure those features to tell what will happen with the weather. The main features that make up the weather are temperature, air pressure, wind speed and direction, humidity, and precipitation.

The lowest level of the atmosphere is called the troposphere. Most of Earth's weather, including the wind and most clouds, is found here.

Temperature is a measurement of how hot or cold the air is. Energy from the sun heats the air around Earth, but it does not reach all of Earth equally or at the same time. This is why some areas are warmer or colder than others.

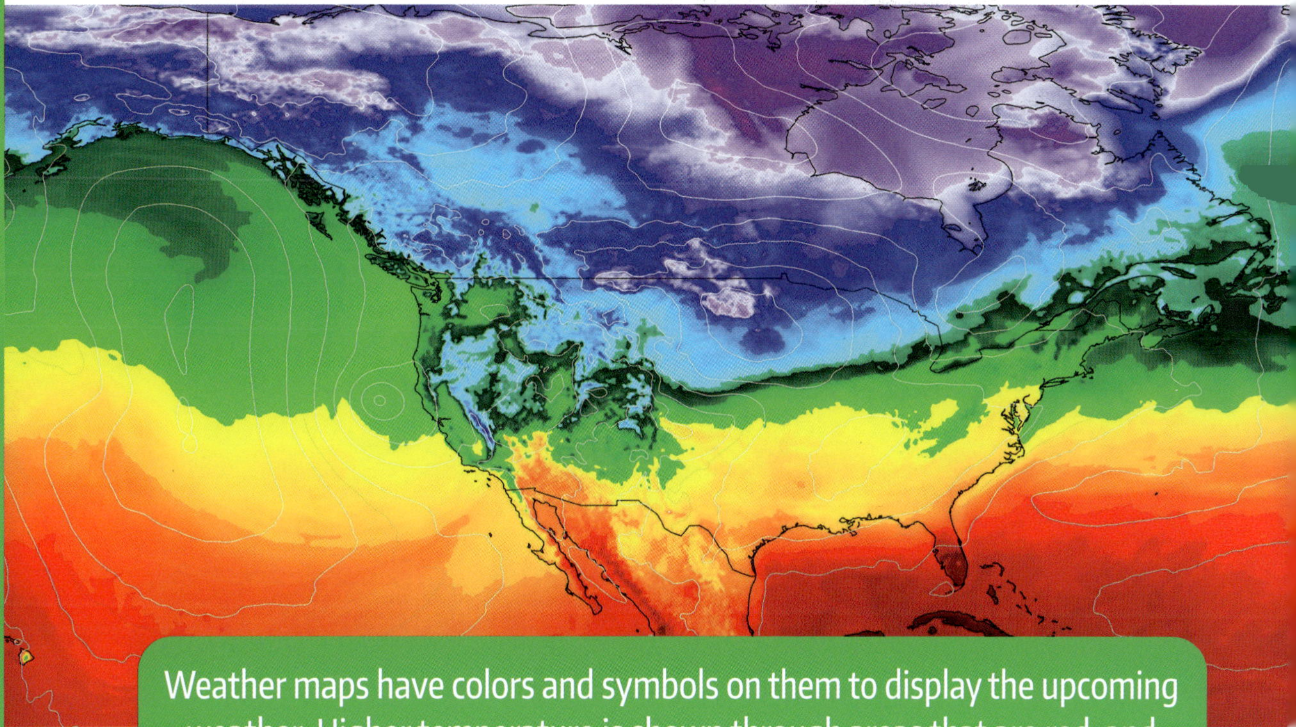

Weather maps have colors and symbols on them to display the upcoming weather. Higher temperature is shown through areas that are red, and lower temperatures are shown through areas that are blue.

compareandcontrast

What color do you think would cover an area like Buffalo, New York, during the winter? What color would cover an area like Orlando, Florida, during the summer?

Air pressure and winds are both impacted by the air's temperature. The gases in the atmosphere have weight that presses down and creates atmospheric, or air, pressure. Warm air weighs less than cold air. As the sun warms the air, the air spreads and moves upward. Colder air around it then moves in. The result is wind. Wind has a great effect on weather.

Water also plays an important part in the weather in several ways. When oceans, lakes, and other bodies of water on Earth heat up, some of the water **evaporates** into the air. The amount of moisture in the air is called humidity. Warm air can hold more moisture than cold air.

When air cools, water vapor begins to turn back into a liquid. The water is in the form of tiny droplets. Many droplets together make clouds. Sometimes the droplets combine and form larger drops. As the drops grow in weight, they eventually fall to Earth as precipitation—such as rain or snow.

Earth's water cycle began around four billion years ago. The water in your cup may have been sipped by dinosaurs!

WORD WISE

THE EVAPORATION OF WATER REFERS TO THE PROCESS OF LIQUID WATER BECOMING A GAS IN THE FORM OF WATER VAPOR. IT OCCURS WHEN HEAT FORCES THE BONDS THAT HOLD WATER MOLECULES TOGETHER TO BREAK.

Way Back When

Many centuries before people had the current knowledge and technology to forecast weather, they had to find ways to predict weather. It was important to know about the weather so they would know what crops to plant and when to plant them.

Some people believed that the gods controlled the weather. Other people looked into the sky and studied what they could see, such as clouds. They looked for patterns so they could know what to expect. Others even studied the behavior of animals. They believed animals' behavior changed before storms or other changes in weather. They recorded their observations in books called **almanacs**.

It is believed animals can sense changes in the weather. This is why your dog may begin acting unusual, such as by hiding under a bed or barking, before a storm.

The Babylonians, who lived in Mesopotamia from the 18th to the 6th century BCE, believed the position of the stars and planets could help them predict weather. They divided the sky into twelve equal parts—called the zodiac—and also studied clouds, winds, and animal behaviors.

In ancient times, the famous Greek thinker Aristotle wrote about his own weather observations. His book on weather was important for more than 2,000 years. Today we know some of his information was inaccurate.

WORD WISE

AN ALMANAC IS A BOOK PUBLISHED ANNUALLY, OR EVERY YEAR, WITH INFORMATION ABOUT WHEN THE SUN WILL SET AND RISE, MOON PHASES, OCEAN TIDES, AND WEATHER FORECASTS.

Creating Tools

People used their own knowledge and estimates to predict weather during ancient times. Later people learned the science of the different weather features. They learned about such things as atmospheric pressure and humidity. They were able to create tools to study these features. These tools include the thermometer to measure temperature, the barometer to measure air pressure, and the hygrometer to measure humidity.

Have you ever felt your ears pop while driving up a mountain or taking off on a plane? This is a result of changes in air pressure, which we measure with a barometer.

Weather balloons are still used today to gather information. They travel with the wind and measure temperature, atmospheric pressure, wind speed, and humidity.

In the 1800s, the telegraph was invented. The telegraph was a device for sending messages over long distances quickly. It allowed scientists to collect and study weather information from many regions on Earth. Other new technology let meteorologists explore the upper atmosphere. They did this by attaching instruments to balloons, kites, and airplanes, which allowed meteorologists to make more accurate forecasts.

Consider This

Why was the telegraph an important invention in the history of forecasting weather? Why is it helpful to study weather from many different places?

How Do We Forecast Weather Today?

Meteorologists today have the assistance of advanced technology and tools to forecast the weather. In addition to the basic tools like thermometers and barometers, they now have radar, satellites, and computers.

Radar technology used by meteorologists sends waves into the atmosphere to find precipitation. When it finds a raindrop or snowflake, the wave reflects back to the radar with **data** that the meteorologists can see and analyze.

Satellites orbiting Earth measure wind speed and direction.

Radar is a system that uses waves of energy to sense objects. It can find a faraway object and tell how fast it is moving. Weather forecasters use it to locate storms and help predict the weather. Satellites are machines sent into outer space to travel around the planet. They take pictures and send back information about the weather conditions happening all around Earth. Scientists take that information and study it. Computers help the scientists see patterns in the data that they gather.

Today computer-drawn maps forecast wind, temperature, and humidity patterns for many levels of the atmosphere. Scientists use mathematical equations—called numerical weather prediction—to calculate high and low temperatures, precipitation, winds, and other weather elements.

WORD WISE

DATA REFERS TO FACTS OR PIECES OF INFORMATION THAT ARE COLLECTED FOR ANALYSIS.

Maps

Weather maps, or forecast maps, are created by meteorologists to display upcoming weather conditions and storms. They use data collected from satellites and other measurement tools to create these maps. Symbols and colors are used on weather maps to show the conditions observed at weather stations. The weather observations used for these maps are collected at thousands of weather stations around the world four times a day. Meteorologists then use this information to predict how the weather will continue changing.

The World Meteorological Organization is a special agency of the United Nations dedicated to providing weather and climate information worldwide. It has 193 members, including countries and territories across the globe.

A storm surge is typically caused by the relationship between wind and the ocean's surface. The water levels rise where the wind is strongest.

An important use of weather maps is to warn people about storms and other dangerous weather. Meteorologists use radar and satellite maps to follow the movement of dangerous weather features. This helps them warn people about approaching bad weather. One of the newest ways to predict danger from the weather is storm surge maps. This type of map alerts those living in coastal areas of potential flooding and hurricanes.

Consider This

Why do you think it would be important for a country or territory to be a member of the World Meteorological Organization?

It is common to see weather fronts displayed on a forecast map. A weather front is a sharp boundary between two different air masses. Air masses are areas of similar types of weather. A cold front is the leading edge of a cold air mass. Cold fronts bring a quick drop in temperature and a fast rise in pressure. They are often accompanied by thunderstorms in summer and snow flurries in winter.

On a weather map, a cold front is shown as a blue line with triangles. The direction in which the triangles point is the direction in which the front is moving. A warm front is the boundary between a cold air mass that is retreating and a mass of warm air. On a weather map, a warm front appears as a red line with semicircles. The semicircles point in the direction the front is moving.

Consider This

Another type of front is called a stationary front, in which a warm front and a cold front meet and stop moving. How might weather maps show stationary fronts? Think of the blue triangles and red semicircles.

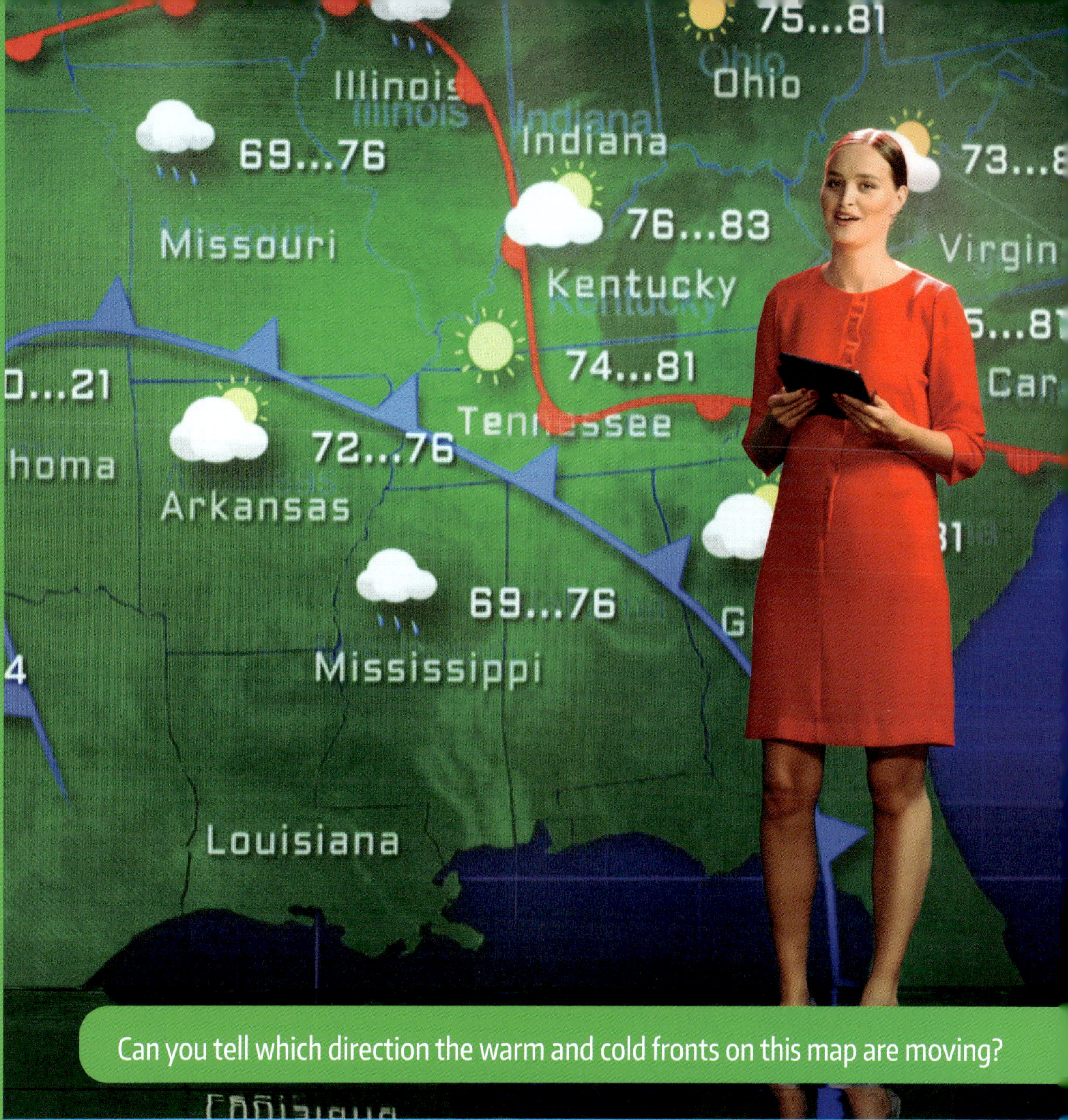

Can you tell which direction the warm and cold fronts on this map are moving?

Flood Watch

Advanced technology and certain tools help meteorologists predict flooding. Tools called rain gauges measure the amount of rainfall. Another tool called a hygrometer measures humidity, or moisture in the air. Most flooding is caused by heavy rains over a long period of time. Even if the rain is not happening where you live, you may be affected by it! The ground can become saturated (full of water). Then the rainwater has nowhere to go. That water can end up in a farmer's fields or in people's homes.

Areas with large populations have a higher risk of flash flooding. The construction of buildings, highways, driveways, and parking lots increases runoff because there is less space for rain to be absorbed by the ground.

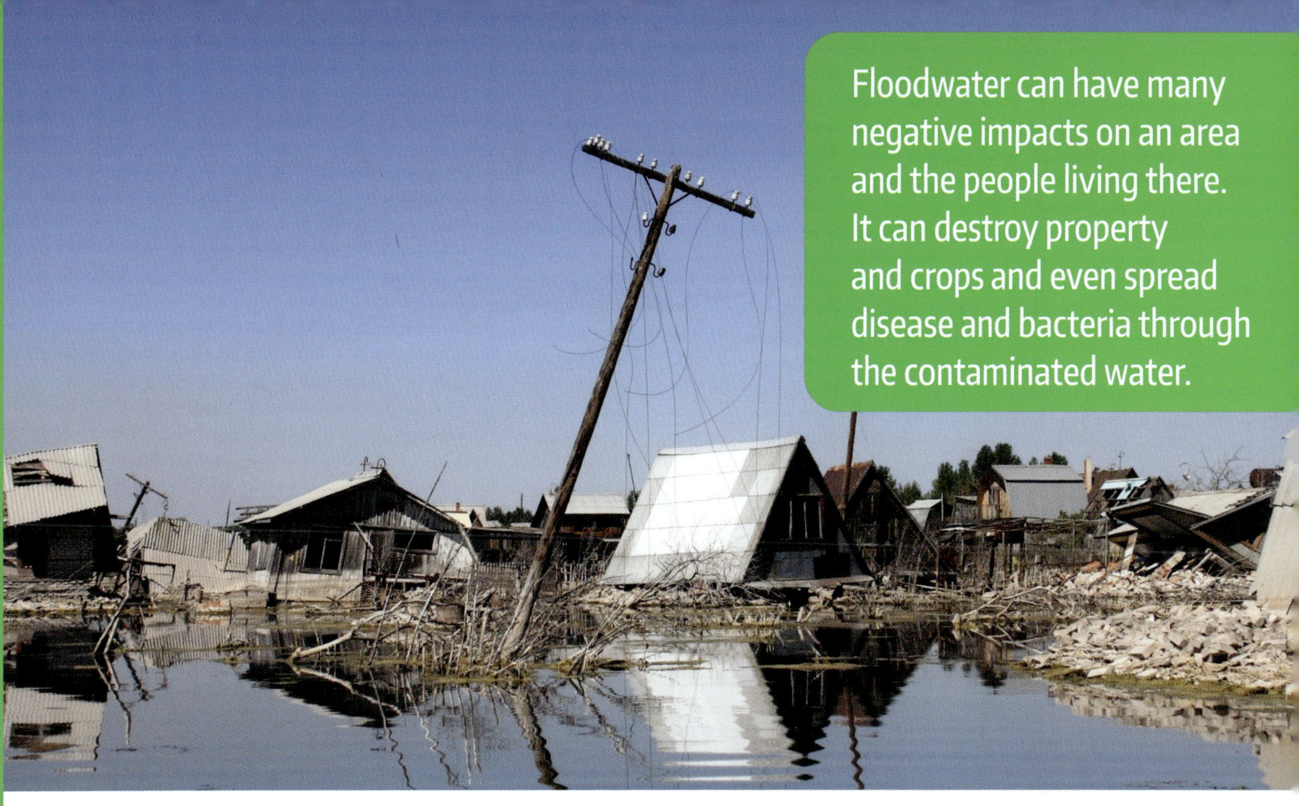

Floodwater can have many negative impacts on an area and the people living there. It can destroy property and crops and even spread disease and bacteria through the contaminated water.

Flooding can be worse if there is not much vegetation, or plants, on the land around rivers. Flash floods are the result of sudden heavy rainfall. They happen quickly and can be deadly. Flash floods can catch people in fast-rising water before they have a chance to escape. A completely dry riverbed can fill up in just a few minutes! Flood watches—as well as flood warnings—can save lives.

Consider This

Why would vegetation around a river impact the strength of a flood?

Spiraling Wind

In order to keep people safe from dangerous storms, meteorologists predict when and where major storms will occur. One storm meteorologists track is called a cyclone. A cyclone is a system of winds that spiral. When such a storm forms over warm oceans close to the equator, it is called a tropical cyclone. Tropical cyclones are also called hurricanes or typhoons in different parts of the world. Their winds swirl at speeds higher than 74 miles (119 km) per hour.

Tropical cyclones are recognizable on radar because they have an "eye," which is a cloudless area of calmer weather in the center of the spiraling winds.

A La Niña event typically follows **El Niño**. La Niña brings stronger winds and makes the water unusually cold from the eastern to the central Pacific.

Meteorologists know what conditions must be present for a cyclone to form. These conditions include certain wind patterns, air pressure measurements, and large ocean patterns like El Niño. When meteorologists know these conditions, they can figure out when cyclones are likely to form. Once a cyclone has formed, it can be tracked using radar. Then, meteorologists can warn people who live in the path of the storm when it will strike and to leave the area or take shelter.

WORD WISE

EL NIÑO REFERS TO THE REOCCURRING CLIMATE PATTERN OF UNUSUAL WARMING OF SURFACE WATERS IN THE EASTERN PACIFIC OCEAN.

Destructive Storms

Although they're more difficult to predict, meteorologists have improved their methods for sensing and forecasting tornadoes. A tornado is a type of storm in which powerful rotating winds form a column of air that reaches down toward the ground. When weather conditions are likely to result in a tornado, meteorologists announce a tornado watch. Tornado winds are the fastest winds on Earth. The winds in a tornado can reach speeds of nearly 300 miles (483 km) per hour.

Tornadoes can be very destructive, and they form quickly. This is why it is important to know that one might be headed your way. In the United States, an agency called the National Weather Service operates a National Severe Storms Forecasting Center. Forecasters there watch the atmosphere for the conditions that can create tornadoes or severe thunderstorms.

Each year, tornadoes cause an average of 70 deaths and 1,500 injuries in the United States.

Winter storms can be powerful as well. Storms featuring low temperatures, strong winds, and large amounts of snow are called snowstorms. The strongest are called blizzards. A blizzard has winds of more than 35 miles (56 km) per hour and visibilities of less than a quarter mile (400 m) for at least three hours.

compare and contrast

How are tornadoes different from cyclones? How are they similar?

Hot and Cold

Long periods of extreme hot or cold weather conditions are known as waves, and they typically affect a large area. In some regions, a cold wave is known as a cold snap or a cold spell. Studying the temperature and air pressure patterns allows scientists to forecast these extreme conditions. Although people may expect warm temperatures in the summer and cold temperatures in the winter, they may not be prepared for extreme temperatures.

Once-green fields full of crops become dead and dry after a **drought**.

Global warming refers to the permanent shifts in temperature as a result of human activity. This coal-fueled power plant releases numerous toxins into the atmosphere, including carbon monoxide and heavy metals.

Extreme heat can lead to drought. Drought can be dangerous to people, and it can ruin crops. Unusually cold temperatures can be dangerous for people as well. If there is warning of a heat or cold wave, people can be prepared to deal with it. Power companies can prepare as well. People tend to use more electricity during hot weather to run air conditioners, while they tend to run heaters in the cold.

WORD WISE

A DROUGHT IS AN EXTENDED PERIOD OF TIME WITHOUT RAIN.

Future Climate

Forecasting the weather is helpful in planning for the upcoming week, but it now has even more possibility. Instead of just forecasting the weather for the next few days, scientists can now look at what the climate will be like over the next few decades. To study climate, scientists look at several conditions, including the greenhouse effect. The greenhouse effect is a warming of Earth's surface and the air above it. It is caused by gases in the air that trap energy from the sun. Without the greenhouse effect, Earth would be too cold for life to exist.

Rising temperatures can cause changes in precipitation levels, storm severity, and sea level.

Excessive heat trapped in Earth's atmosphere negatively impacts many ecosystems and the species within them. Coral has become one of the world's most quickly declining species due to rising ocean temperatures.

Scientists have discovered that humans are increasing the greenhouse effect through their actions. When people drive cars, work in factories, or use electricity at home, they burn coal, oil, and other fuels. Burning these **fossil fuels** adds extra gases to the air. The extra gases trap more heat. Scientists say that this has led to global warming. Forecasters are continuing to work hard to predict the effect global warming will have on nature and Earth's resources in the coming years.

WORD WISE

FOSSIL FUELS ARE FORMED FROM THE REMAINS OF LIVING ORGANISMS, AND THEY INCLUDE COAL, OIL, AND NATURAL GAS.

Glossary

atmospheric pressure The weight of the air; also known as barometric pressure.

broadcast To air on television or radio.

contaminte To make something dirty or polluted.

ecosystem A community of living things interacting with their environment especially under natural conditions.

estimate An educated guess.

forecast To calculate or guess that something will happen in the future based on data from the past.

inaccurate Not correct.

invention A new creation.

predictable Able to be foretold.

recognizable Able to be identified.

runoff An unconfined flow of water over the ground.

severity The seriousness of something.

spiral Winding or circling around a center.

storm surge The rising of the sea as a result of changes in atmospheric pressure and wind.

For More Information

Books

McCoy, Erin L. *Global Warming*. Buffalo, NY: Cavendish Square, 2019.

Nelson, Penelope. *Forecasting Weather*. Minneapolis, MN: Cody Koala, 2019.

Rathburn, Betsy. *Tornadoes*. Minneapolis, MN: Bellwether Media, 2020.

Websites

National Geographic: Climate Change
education.nationalgeographic.org/resource/climate-change/
Read more about the ongoing crisis of global warming.

National Geographic: Meteorology
education.nationalgeographic.org/resource/meteorology/
Learn more about the study and practice of meteorology.

National Geographic: Storms
nationalgeographic.co.uk/storms
Explore helpful articles and recent headlines about storms.

Index